NIST Technical Note 1641

PASS Sound Muffle Tests Using A Structural Firefighter Protective Ensemble Method

J. Randall Lawson
Fire Research Division
Building and Fire Research Laboratory
National Institute of Standards and Technology
Gaithersburg, MD 20899-8661

July 2009

Department of Homeland Security
Janet Napolitano, Secretary

Federal Emergency Management Agency
W. Craig Fugate, Administrator

United States Fire Administration
Glenn Gaines, Assistant Administrator

U.S. Department of Commerce
Gary Locke, Secretary

National Institute of Standards and Technology
Patrick D. Gallagher, Deputy Director

Certain commercial entities, equipment, or materials may be identified in this document in order to describe an experimental procedure or concept adequately. Such identification is not intended to imply recommendation or endorsement by the National Institute of Standards and Technology, nor is it intended to imply that the entities, materials, or equipment are necessarily the best available for the purpose.

CONTENTS

List of Figures ii

List of Tables ii

Abstract 1

1.0 INTRODUCTION 2

2.0 TEST DESCRIPTION 2

    2.1 TEST SPECIMENS 2

    2.2 TEST INSTRUMENTATION 3

    2.3 TEST ENVIRONMENT 4

    2.4 TEST PROCEDURE 5

        2.4.1 PASS DEVICE TESTING 5

        2.4.2 ORDER OF TESTING 6

        2.4.3 DETAILED TEST PROTOCOL 10

3.0 TEST RESULTS AND DISCUSSION 12

4.0 RECOMMENDATIONS FOR FUTURE WORK 22

5.0 CONCLUSIONS 23

6.0 ACKNOWLEDGEMENTS 23

LIST OF FIGURES

Figure 1 Characterization of the dBA decibel scale 3
Figure 2 Photograph showing calibration of sound meter and laser
used for alignment of sound meters and PASS devices 4
Figure 3 Photograph showing room and arrangement for unobstructed
testing of clip-on PASS devices 5
Figure 4 Diagram showing standard arrangement for PASS sound level testing 6
Figure 5 Photograph showing SCBA with integrated PASS on test stand 7
Figure 6 NFPA 1982 position 1 8
Figure 7 NFPA 1982 position 2 8
Figure 8 NFPA 1982 position 3 9
Figure 9 NFPA 1982 position 4 9
Figure 10 NFPA 1982 position 5 10
Figure 11 PASS Device A – Change in muffle position sound level vs.
unobstructed sound level (0 line) 14
Figure 12 PASS Device B – Change in muffle position sound level vs.
unobstructed sound level (0 line) 15
Figure 13 PASS Device C – Change in muffle position sound level vs.
unobstructed sound level (0 line) 16
Figure 14 PASS Device D – Change in muffle position sound level vs.
unobstructed sound level (0 line) 17
Figure 15 PASS Device E – Change in muffle position sound level vs.
unobstructed sound level (0 line) 18
Figure 16 Photographs showing background sound level measurement tests 22

LIST OF TABLES

Table 1 Summary of PASS Device Sound Level Variability 19
Table 2 Fire Apparatus and Fire Fighting Equipment Sound Levels 20
Table 3 Background Sound Levels 21

PASS Sound Muffle Tests Using A Structural Firefighter Protective Ensemble Method

By

J. Randall Lawson

Abstract

Firefighters and other emergency responders often work in adverse environments. The operating environments can be very noisy. Personal Alert Safety Systems (PASS) devices are safety systems that emit an audible alarm signal when an emergency responder stops moving. This alarm signal serves as a means for others to rapidly locate a responder who may be injured or incapacitated. The 2007 edition of National Fire Protection Association, NFPA 1982, Standard on Personal Alert Safety Systems (PASS) established a means to evaluate the muffling (attenuation) of sound level from a PASS device worn by an emergency responder who is down on the floor. The standard prescribes five different test positions that are assumed when measuring sound level attenuation from a firefighter, fully dressed in their NFPA 1971 fire fighting ensemble, while wearing a PASS device that is in full alarm. NIST tested five different pass devices in an open room environment to determine relative signal level attenuation for each device. These tests demonstrate sound level losses ranging from approximately 9 % dBA to 19 % dBA. These signal losses are associated with the type of device, direction of measurement relative to the orientation of the PASS device, and attenuation from the firefighter's body obstructing the sound path. This report also contains background sound level measurements with dBA measurements from fire fighting apparatus and power tools that are commonly used on the fireground.

KEY WORDS: Alarms; emergency responder; firefighter; muffle; PASS; safety; sound attenuation; sound level; standards

1.0 INTRODUCTION

Personal Alert Safety Systems (PASS) devices have been used by the fire service since or before the early 1980's for locating and rescuing downed firefighters. The National Fire Protection Association (NFPA) published the first standard for PASS devices, in 1982[1]. These devices electronically sense motion and the lack of motion from a person wearing the device. When the device senses that there has not been any motion, after a prescribed period of time, the device goes into alarm by emitting an audible signal. These devices generally have lights that also flash when the alarm is activated. The combination of the audible alarm and flashing lights are designed to assist other emergency responders in locating a potentially injured firefighter. PASS devices are also available with a radio transmitter that sends the alarm to a command post when it is activated. With all of these PASS devices the emergency alarm can also be activated manually by the user.

The objective of this study was to better understand the relative change in sound level from an activated PASS when it is being used by a firefighter. The alarm signal sound level can be affected by a range of factors including location of PASS device on the firefighter (front, side, or rear), position of the firefighter (prone, supine, fetal), the volume of the room, the wall linings or materials, the presence of furniture or other objects, and background sounds. NFPA PASS Standard 1982 (2007) includes a muffle test where sound attenuation is measured in a sound chamber. While this allows each PASS device to be evaluated under carefully controlled conditions, the testing protocol includes only the location of the pass and the position of the firefighter. This study examined the attenuation of the alarm signal in a more typical room environment than a sound chamber. This study also documented typical background sound levels for fire apparatus, fire fighting equipment, offices, living room, and vehicles. Comparing the background sound levels with the sound attenuation data provides a better understanding as to whether the decreased sound levels will be distinguishable above background sounds. Figure 1 provides a comparison of various sound levels on the standard dBA decibel scale [2, 3]

2.0 TEST DESCRIPTION

2.1 TEST SPECIMENS

Five different PASS devices meeting the requirements of the 1998 edition of NFPA 1982 were evaluated in this study [4]. Two PASS devices were clip-on designs and three were integrated into the construction of Self-Contained Breathing Apparatus (SCBA) systems. The PASS designs that were clip-on are stand alone units that do not depend on the operation of the SCBA or other firefighter equipment. The clip-on devices are normally attached by a clip to the front of a firefighter's SCBA harness. One of the integrated

PASS systems had the alarm signal device located on the back of the SCBA pack, and all

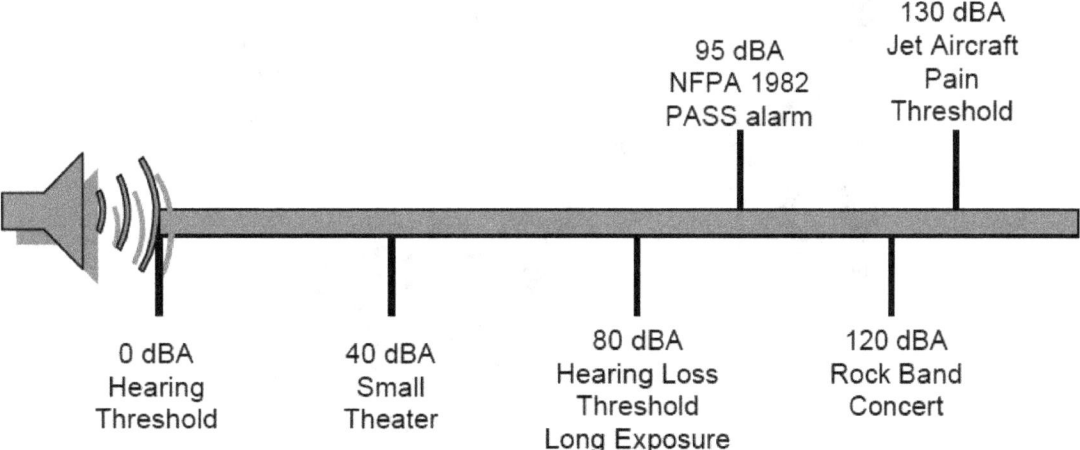

Figure 1. Characterization of the dBA decibel scale.

other integrated PASS alarm signal emitters were located on the front straps of the harness. At the time of testing, all five PASS device designs were typical of devices used by emergency responders. Four of the devices were new, and one clip-on device had been previously used in nondestructive testing, at NIST.

The two clip-on PASS devices were activated by manually removing a key from the device. The three integrated PASS device systems were connected into the SCBA air management system and were activated by SCBA operation. All five devices were battery powered, and they all had a switch that could be manually operated for activating the alarm signal.

2.2 TEST INSTRUMENTATION

The sound level meters used during this study were EXTECH Model 407764 data logging meters with the capability of storing up to 16,000 readings.* These meters recorded the sound level and a timestamp. Each handheld meter utilized a RS-232 port to download logged data. Each of the sound level meters was initially calibrated by the manufacturer using a NIST traceable sound level standard. Additionally, before each PASS device test, the sound level meters was rechecked using a portable sound level calibrator. This battery operated NIST traceable sound level calibrator provides calibration output signals at either 94 dBA ±0.5 dBA or 114 dBA ±0.8 dBA. Since the primary decibel level for PASS output is 95 dBA, all sound level calibration checks were carried out at the 94 dBA level setting (Figure 2).

*Certain commercial entities, equipment, instruments, products, or materials are identified in this document in order to describe a procedure or concept adequately or to trace the history of the procedure and practices used. Such identification is not intended to imply recommendation, endorsement, or implication that the entities, products, materials, or equipment are necessarily the best available for the purpose. Nor does such identification imply a finding or fault or negligence by the National Institute of Standards and Technology.

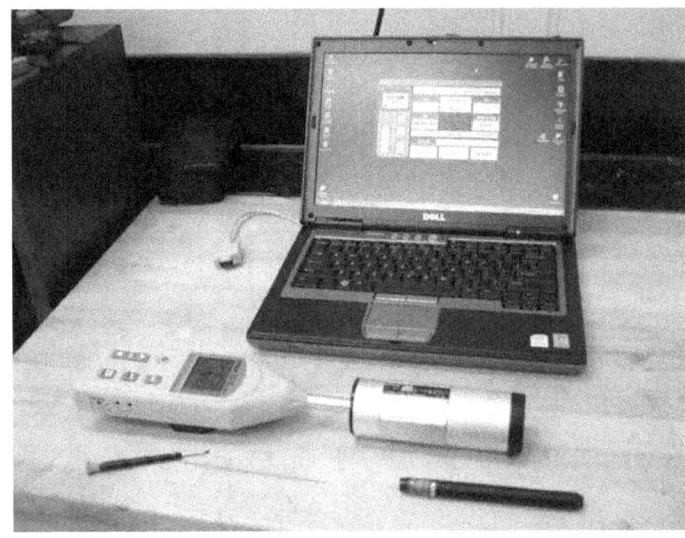

Figure 2. Photograph showing calibration of sound meter and laser used for alignment of sound meters and PASS devices.

2.2.1 Measurement Uncertainty

The manufacturer's published accuracy for the sound level meters over a range of 30 dBA to 130 dBA is ±1.5 dBA with a resolution of 0.1dBA. Although the specifications for the sound level meters and the calibration standard are relatively tight, it is estimated that the combined standard uncertainty (i.e., estimated standard deviation) u_c for these sound level measurements is on the order of ±6 dBA and a coverage factor $k = 2$. Since it can be assumed that the possible estimated values of the standard are approximately normally distributed with an approximate standard deviation u_c, the unknown value of the standard is believed to lie in the interval defined by $U = k\,u_c$ with a level of confidence of approximately 95 %.

2.3 TEST ENVIRONMENT

2.3.1 TEST ROOM

The test room used for this study was similar to a space that would be found in an office building or a light industrial building. The average background sound level measured in this test room was (42 ±2) dBA which is similar to that found in a typical office environment. The test room used in this study measured 4.1 m (13 ft 6 in) high, 4.8 m (16 ft) wide, and 6.5 m (21 ft 6 in) long. The room was clear of any obstructions, including furniture and columns. The upper parts of the walls were covered with 50.8 mm (2 in) thick, fabric covered, low density, glass fiber batting. The lower part of the walls was covered with painted steel panels and some wall surfaces extended into the room to provide surfaces for utility outlets. The room had a concrete ceiling with a textured spray paint finish, and the concrete floor was covered with vinyl flooring tile. See photograph of the room test area in Figure 3. The ceiling and the walls also had exposed air-conditioning ducts and outlets. The room's temperature and humidity were controlled by the building's air-conditioning system. Room temperature was nominally (21 ± 2) °C (70 ± 4) °F and the relative humidity was typically (50 ± 2) %. Temperature and humidity conditions are provided since these environmental conditions may have an affect sound propagation.

Figure 3. Photograph showing room and arrangement for unobstructed testing of stand alone PASS devices.

2.3.2 BACKGROUND SOUND LEVEL MEASUREMENTS

The background sound level measurements were made inside buildings or vehicles to obtain typical background sound levels that may be found inside residential and commercial office buildings or inside a fire department officer's vehicle. The fireground fire apparatus and equipment sound level measurements were all made outside.

2.4 TEST PROCEDURE

2.4.1 PASS DEVICE TESTING

The test procedure was similar to the muffle test as described by the NFPA 1982 Standard on Personal Alert Safety Systems (PASS), 2007 edition [1]. For conditioning purposes, all PASS devices were maintained in a nominal (21 ±2) °C (70 ± 4) °F and 50 ±2 % relative humidity environment for a period greater than 30 days prior to being tested. The basic test layout from the NFPA standard is shown in Figure 4. As noted above in section 2.3, the room obtained for these tests was not wide enough to allow for all four measurements (0°, 90°, 180° and 270°) to be made without repositioning. As a result, one set of sound measurements was made to obtain data for the 0° and 180 ° positions (↑) of the PASS device. (Figure 4) Then the PASS device was turned 90° counterclockwise (←) so that sound instruments could measure the second set of data and obtained readings for the 270° and 90° locations. (Figure 4)

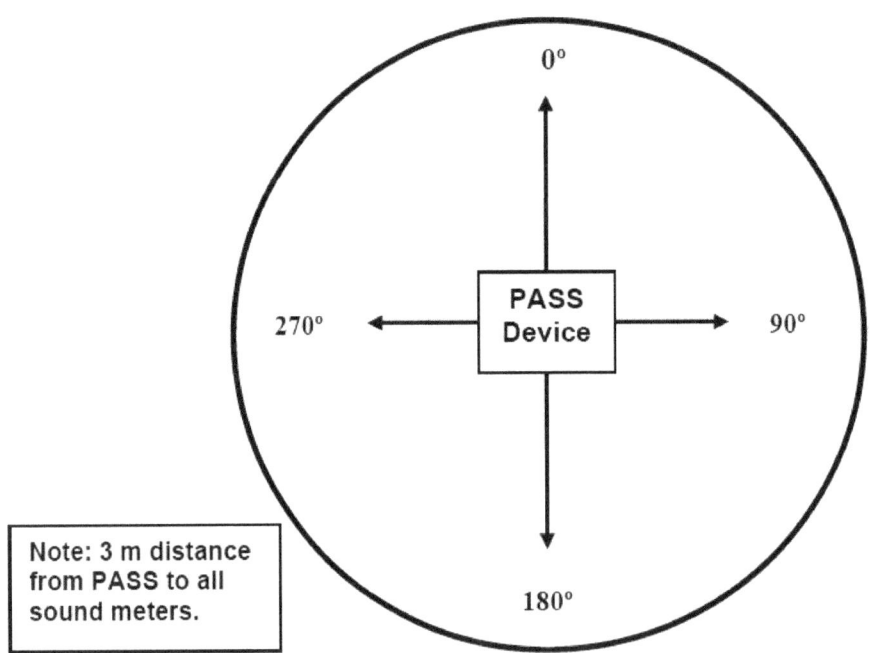

Figure 4 Diagram showing standard arrangement for PASS sound level testing.

Before any sound level tests were conducted, new batteries were placed in each PASS device, and the device was briefly activated to insure that it was operating properly. Additionally, all sound meters were operated with fully charged batteries.

2.4.2 ORDER OF TESTING

The following describes the testing order used for measuring sound muffling for each of the five PASS devices: Photographs showing each of the test positions are located in Figures 6 through 10. In the photographs, the direction of the firefighter's head is always pointing toward the 0° measurement location.

1. Each device was tested when suspended from a support so that no obstructions existed between the device and the sound level meters. This was done to measure maximum baseline sound level output for each PASS for comparison with results from the muffled tests. (Figures 5 through 10)

2. Position 1: Face down with arms fully extended out to the sides. (Figure 6)

3. Position 2: Supine left as far as possible, arms down along sides (Figure 7)

4. Position 3: Supine right as far as possible, arms down along sides. (Figure 8)

5. Position 4: Fetal, knees drawn to chest as far as possible, arms around legs, and lying on right side. (Figure 9)

6. Position 5: Fetal, knees drawn to chest as far as possible, arms around legs, and lying on left side. (Figure 10)

Note that in positions 4 and 5 that fire fighters could not wrap their arms around their legs in a fetal position. (Figures 9 and 10)

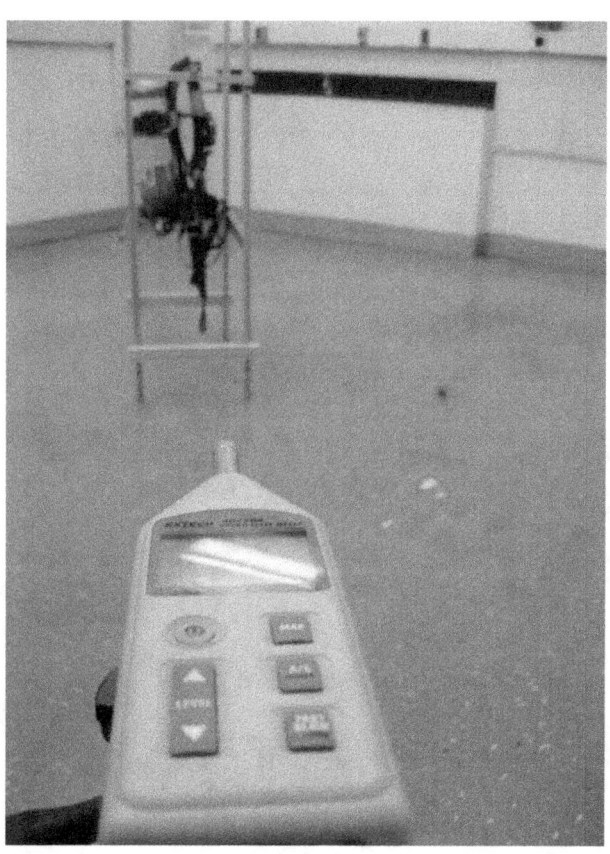

Figure 5. Photograph showing SCBA with integrated PASS on test stand.

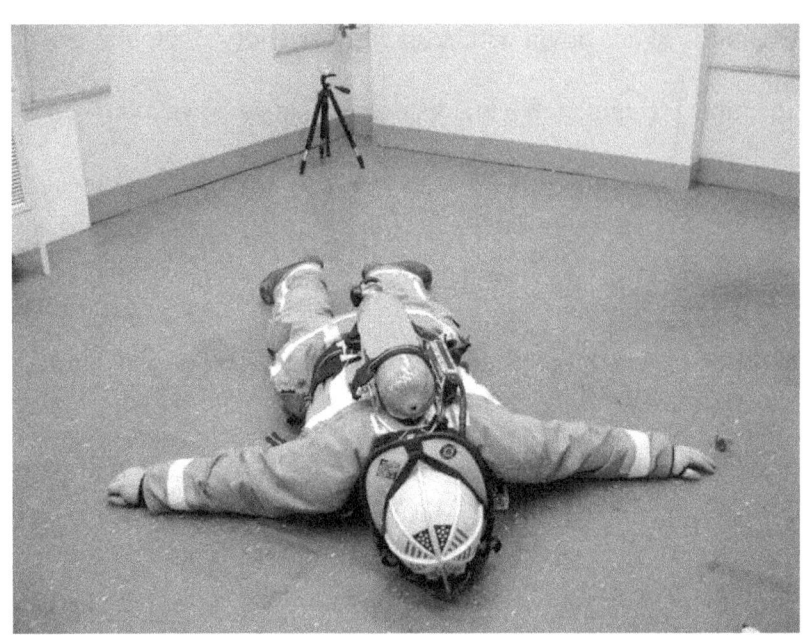

Figure 6. Position 1: Face down.

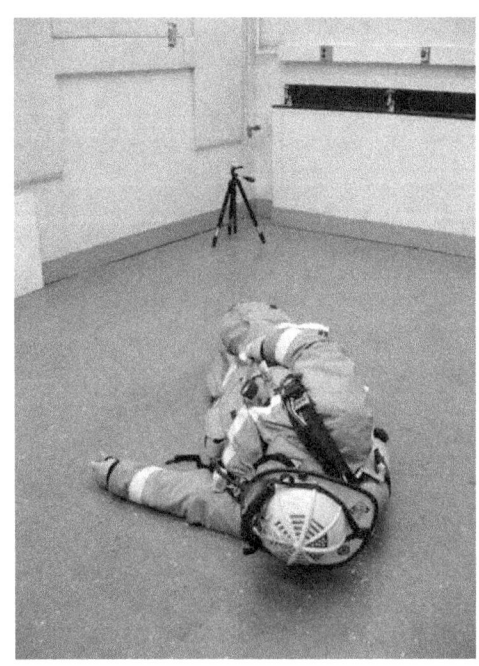

Figure 7. Position 2: Supine Left.

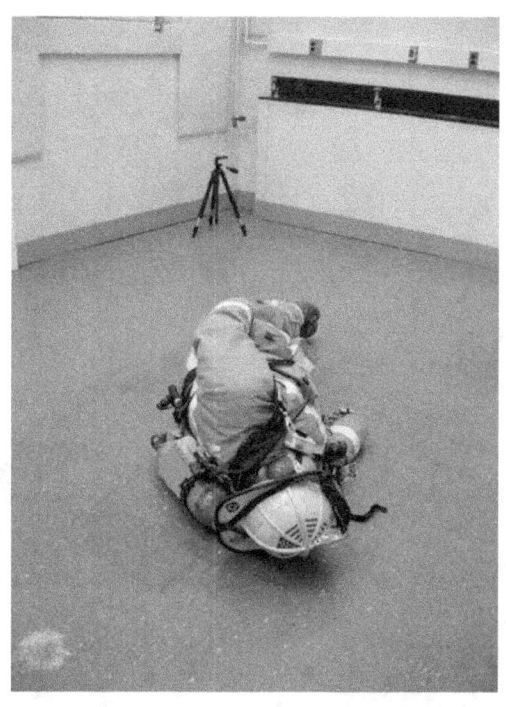

Figure 8. Position 3: Supine right.

Figure 9. Position 4: Fetal right.

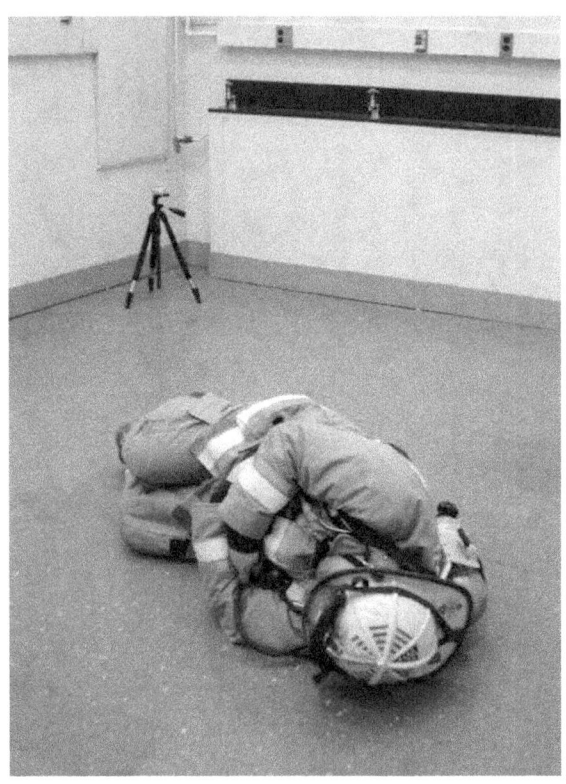

Figure 10. Position 5: Fetal left.

2.4.3 DETAILED TEST PROTOCOL

The following protocol was used with all of the PASS tests.

2.4.3.1. Instrumentation Layout

The center of the test floor was marked by placing a black tape X on the point of symmetry. Then two other points were measured 3 m (9.9 ft.) in opposite directions from the center and marked on the floor. This became the 0° sound measurement point and the 180° sound measurement point. By rotating the PASS device 90° these sound measurement points also became the 90° sound measurement point and the 270° sound measurement points. Each of these three points was also marked with an indelible marking pen. Alignment of the three points was accomplished by using a laser beam across the floor.

Before each set of tests, each of the sound meters was calibrated. The sound meters were then attached to camera tripods and were positioned to align the meter's microphone directly above the previously described measurement locations. At this point in the test, the sound meters were operating but were not recording data. The height of the sound meters above the floor was set at 0.6 m (2 ft.). This was chosen to approximate the head height of a crawling firefighter.

2.4.3.2 Supports for Unobstructed Tests

The unobstructed tests were conducted by suspending PASS devices over the point of symmetry from vertical supports. The stand alone devices were suspended from a laboratory stand, and the integrated devices were suspended from an aluminum frame that allowed for the SCBAs to be suspended by the shoulder straps. (Figures 3 and 5)

2.4.3.3 Firefighter Ensemble Tests

When sound muffling tests were conducted using the firefighter ensemble, the sound producing part of the PASS device was located over the symmetry point on the floor. In these tests, the firefighter wore their normal structural fire fighting ensemble that meets NFPA 1971 Standard on Protective Ensembles for Structural Fire Fighting and Proximity Fire Fighting [1]. (Figures 6 through 10)

2.4.3.4 Conducting the Test

After all elements for a test were in place and operating, each sound level meter was set to begin logging, and then the PASS device was manually activated. The PASS device remained in a full alarm mode for a period of approximately two minutes while the instruments collected sound level data. At the end of the two minute period the PASS device was silenced and data logging was stopped. The PASS device was then turned 90°, and the procedure was repeated. All test personnel wore hearing protection during the periods of alarm activation.

2.4.3.5 Managing the Sound Level Data

When the set of two tests were completed (which allowed for taking measurements at the 0°, 90°, 180°, and 270° points) the sound meters were attached to a portable computer and each data file was downloaded, saved, and archived. The data were then plotted, analyzed, and are presented in Section 3.

2.4.4 BACKGROUND SOUND LEVEL MEASUREMENTS

Measurements for background sound levels were conducted using similar methodology as that for the PASS sound measurements discussed above. Two different fire department companies participated in this measurement effort (see acknowledgements), and it involved more than twenty pieces of fire department equipment that are described in Tables 2 and 3, of Section 3.2. All of these sound level measurements were made outside at each of the respective firehouses. The measurements were made on clear days with the temperature at about 20 °C (68 °F). Sound level measurements were made concurrently with two calibrated sound meters located 0.6 m (2 ft.) above ground level. One measurement was typically made 3 m (9.9 ft.) from the front of the apparatus or equipment and is designated the 0° measurement. The second measurement was typically made 3 m (9.9 ft.) from the side of the apparatus or equipment and is designated the 90° measurement. Additionally, the 90° sound measurements were typically made at locations on the exhaust side of the equipment where the highest sound levels would be expected. The radio sound measurements were made at 1 m (3.3 ft.) from the radio at both the 0° and 90° locations relative to the speaker/mic position. The dBA values for each of the background sound level measurements represent an average observed over a 30 second measurement time period. Results are presented in Section 3.2.

Additional sound measurements were also made to quantify some normal sound levels that may be experienced in residential, office, and building mechanical space settings. All of these measurements, except measurements on the inside of the vehicle, were made at a location 0.6 m (2 ft.) above ground level. These measurements were made at only the 0° location relative to the item of interest and did not include a measurement at a 90° location. These results are also reported in section 3.2

3.0 TEST RESULTS AND DISCUSSION

3.1 PASS DEVICE RESULTS

Test data for each of the PASS devices are presented in Figures 11 through 15. There are four plots for each PASS device. Each of the four plots represents one of the sound measurement points. The "0" sound level line on each plot represents the reference sound level output from the unobstructed PASS device. The decibel level bar graph data plots extending from the "0" line represents the change in sound level relative to the unobstructed sound level test condition. The bar graph plots show respectively from left to right results from each ensemble muffle position 1, 2, 3, 4, and 5. The PASS device designation is located on the top of each plot and the figure title. (Figures 11 through 15) Each graphed data value incorporates a reference bar showing the sound test measurement uncertainty of ± 6 dBA.

The objective of these data plots and values reported in the summary table (Table 1) is to provide the reader with a comparison of sound level muffling as it relates to the position of the PASS device relative to direction of sound travel away from the device. From these data plots, it is observed that sound levels for all five PASS device systems changed with ensemble test position. Note that all PASS alarm signal devices were located on the front harness, except for PASS device D which was located on the back. This alarm signal device location on the back of the harness produced a more uniform distribution of sound from the PASS system. In some cases, the sound level did not fall below the unobstructed test level but increased above this baseline. Each of the PASS devices, during ensemble testing, exhibited cases where measured sound levels exceeded the unobstructed test baseline values. This appeared primarily to be a result of the PASS device alarm orientation relative to the sound meters. In the unobstructed tests the PASS devices were suspended vertically. In this orientation, some PASS device alarm components were directed towards the floor. In the ensemble tests the PASS devices typically were laying horizontally, changing the output direction from the alarm and in some cases possibly providing a more direct signal path to the sound meters.

PASS device B exhibits the greatest decrease in sound level during the ensemble muffling tests. It shows a sound level amplitude reduction of approximately 15 dBA in the 0°, ensemble position 5 test and approximately 19 dBA decrease in signal amplitude with the 180°, ensemble position 5 test condition, Figure 12. Conversely, PASS device D exhibits sound level increases for three of the five 90° ensemble positions. Also, D shows only slight changes in signal output for the 0° and 180° positions. (Figure 14)

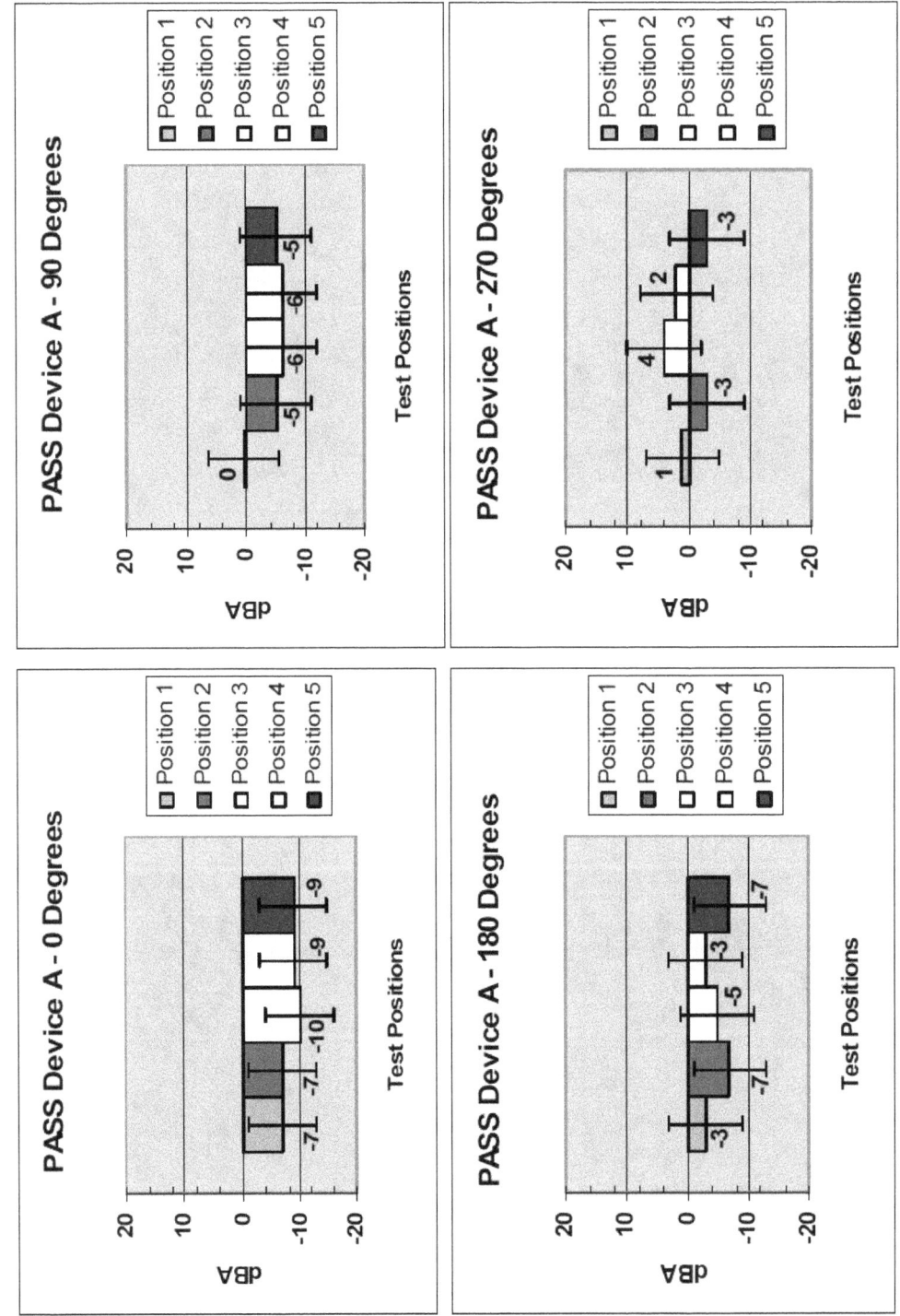

Figure 11. PASS Device A – Change in muffle position sound level vs. unobstructed sound level (0 line).

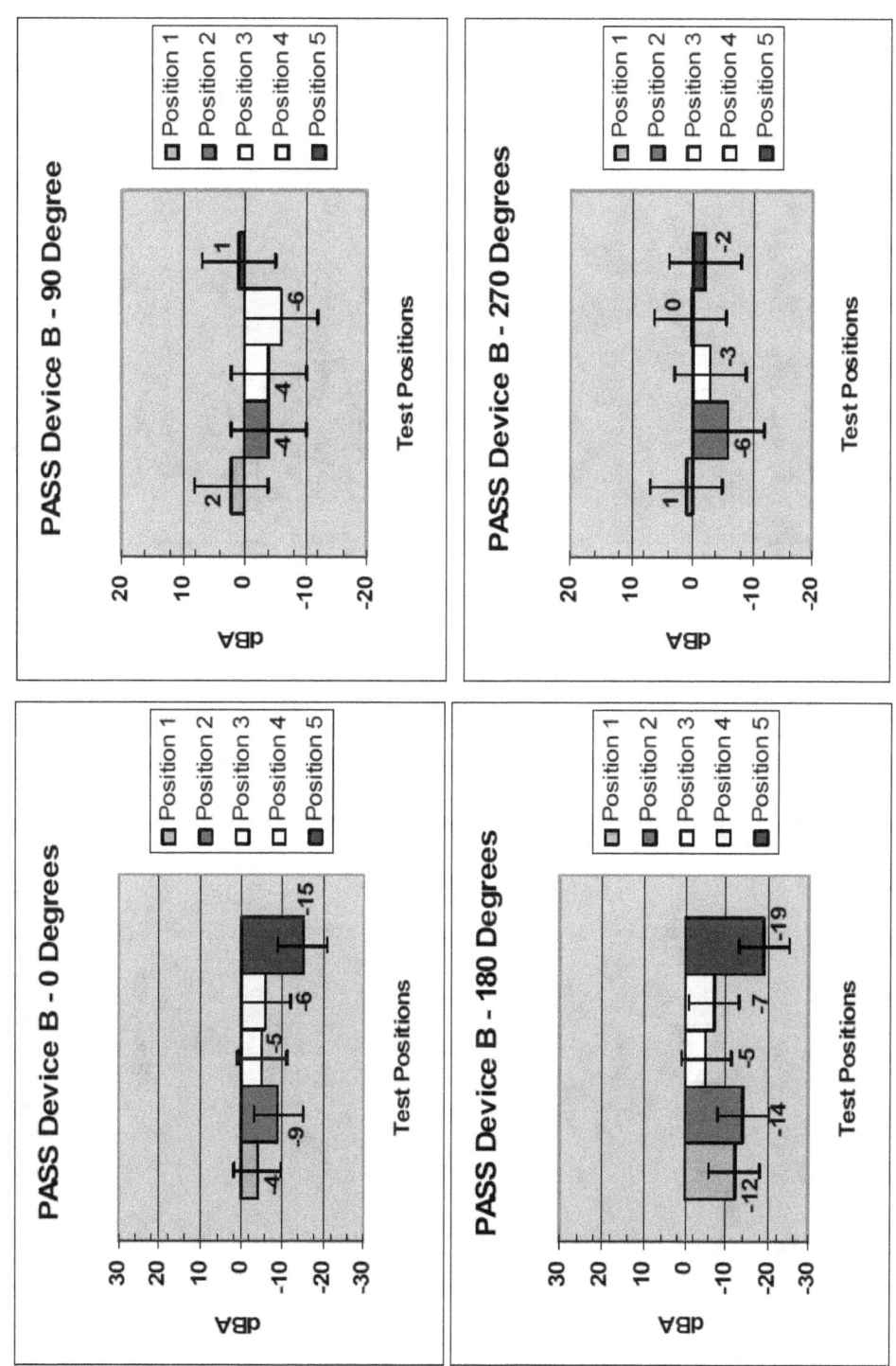

Figure 12. PASS Device B – Change in muffle position sound level vs. unobstructed sound level (0 line).

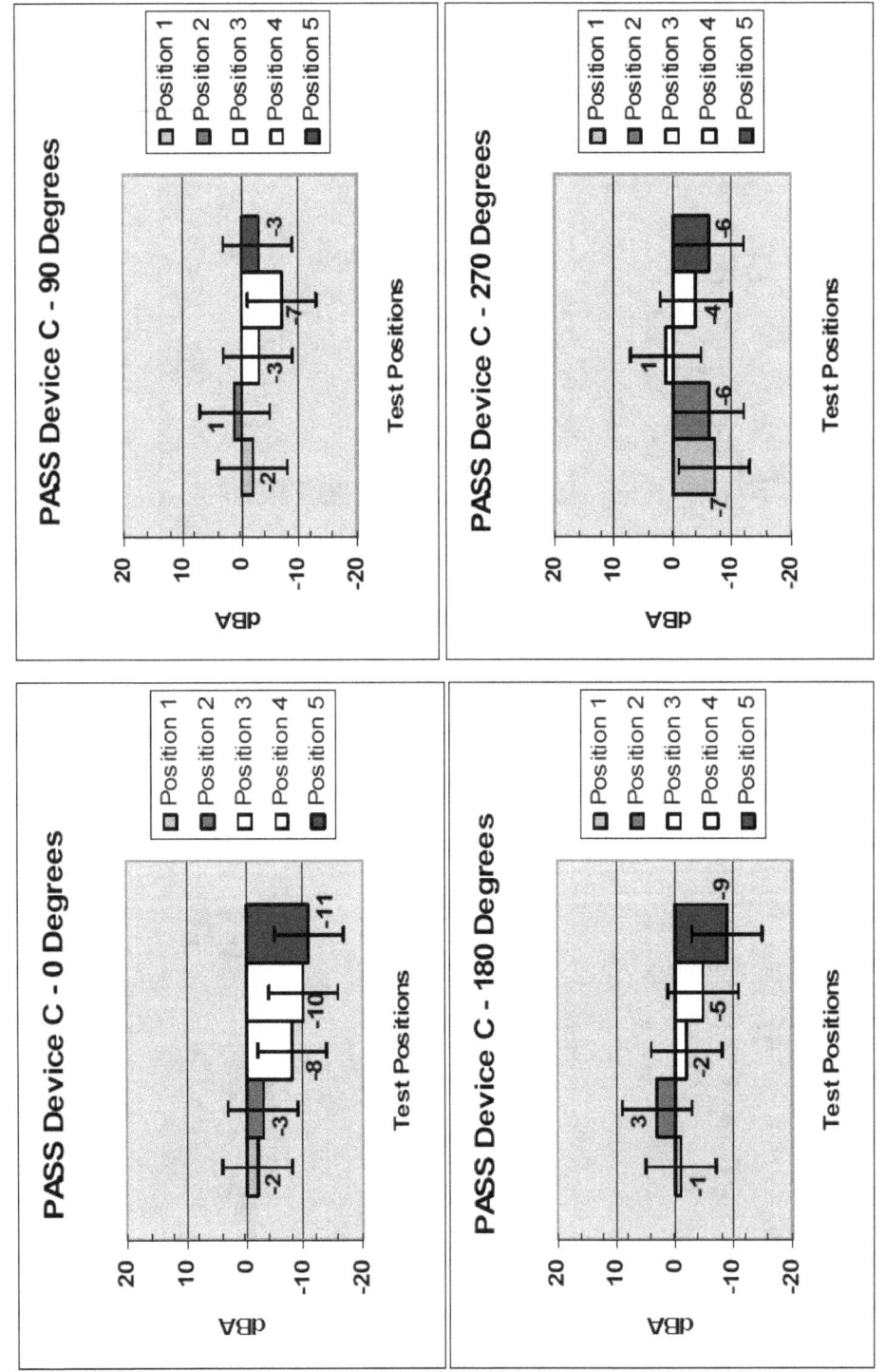

Figure 13. PASS Device C – Change in muffle position sound level vs. unobstructed sound level (0 line).

A summary of the data plots is located in Table 1. The objective of this table is to provide the reader with a summary of PASS device results focused on maximum changes measured for sound levels from each device; additionally, the overall dBA percent variation is shown. Note that the sound level changed over a range of approximately 9 % to 19 % for the set of PASS devices tested. This amount of variation is not small when it is considered that dBA sound level measurements are made on a logarithmic scale. Therefore, the change in apparent sound level to the human ear could be significant.

Table 1
Summary of PASS Device Sound Level Variability

Note: Per section 2.2.1, measurement uncertainty for all table values is ±6 dBA.

PASS Device	Maximum Difference (dBA)	Minimum Difference (dBA)	Percent Variation (averaged over 5 positions) (% dBA)
A	-10	0	14.1
B	-19	0	19.2
C	-11	1	14.0
D	+8	0	13.2
E	-6	0	8.6

An additional observation related to testing, note in Figures 9 and 10 that NFPA 1982 uses the following definitions for the positions: Position 4 Fetal, knees drawn to chest as far as possible, arms around legs, and lying on right side. Position 5 Fetal, knees drawn to chest as far as possible, arms around legs, and lying on left side.

As clearly observed in Figure 10, the test subject could not pull the legs up high enough to achieve the prescribed position. It was found that the firefighting ensembles restricted body movement preventing the firefighter from bringing the legs up high enough to wrap the arms around their legs.

3.2 BACKGROUND SOUND LEVEL RESULTS

Table 2 provides information on background sound levels that may be experienced on the fireground. The primary focus of this table is to provide data on sound levels produced by fire department apparatus and fire fighting equipment. It should be noted that results in this table represent sound measurements for single pieces of equipment. The results do not account for sound levels that may actually be found on the fireground when multiple fire apparatus and multiple pieces of fire fighting equipment are operating at the same time. The sound levels reported in Table 2 represent the mean value from sound level sampling during a time period of approximately 30 seconds. Each of these reported values has a measurement uncertainty of about ± 6 dBA. The first row of data provides background sound levels measured at each of the two test sites. Generally, these ambient background sound levels were not directionality different and only single values are reported for these cases. The first column provides a description of the equipment being tested. The second column identifies the distance between the measurement instruments and the test item. The third and fourth columns provide 0° and 90° dBA values measured with Company 12 equipment and the fifth and sixth columns provide 0° and 90° dBA values measured with Company 53 equipment. Column seven provides comments designed to assist the reader with understanding conditions for each of the measurements.

Table 2
Fire Apparatus and Fire Fighting Equipment Sound Levels

Note: Per section 2.2.1, measurement uncertainty for all table values is ±6 dBA.

Description	Distance	Company 12		Company 53		Comments
	(m)	0° (dBA)	90° (dBA)	0° (dBA)	90° (dBA)	
Test Site	n/a	62	62	54	54	Ambient sound level.
Fire Engine	3	80	79	77	76	Engine at Idle
	3	87	85	81	80	Engine high Idle
Siren	3	109	100	109	104	
Federal Q, Electronic	3	110	108	108	109	
Air Horn	3	109	107	109	106	
Apparatus Backup Alarm	3	81	83	90	87	
Operating Pump	3	85	85	79	80	Idle
	3	97	95	88	91	Max RPM
PPV Fan, Gas Engine	3	106	97	97	92	
Chain Saw		81	80	75	71	Idle
	3	100	96	106	102	Max RPM
Circular, Cutoff Saw	3	81	79	75	71	Idle
	3	105	103	106	102	Max RPM
Portable Radio, High Volume	1	79	76	83	75	Without Apparatus running
	1	87	85	85	82	With Apparatus running

Table 3 provides 30 second average sound level measurement values for a variety of residential, office, vehicle, and mechanical room locations. The table provides information on the location or type of equipment being examined, distance from the equipment, the measure sound level in dBA, and comments describing the measurement. Additionally, smoke detector and commercial fire alarm dBA values are provided for comparison. These data are based on the NFPA 72 standard [5] for sound levels for residential smoke alarms and commercial fire alarms.

Table 3
Background Sound Levels

Note: Per section 2.2.1, measurement uncertainty for all table values is ±6 dBA.

Location and Equipment	Distance (m)	(dBA)	Comments
Two Person Office	n/a	43	Measurement center of office space (3.3 m W x 4.8 m L x 3.3 m H)
Ringing Office Telephone	2	65	Same office describe above.
Home Living Room	n/a	36	Measurement made in center of room, doors and windows closed
Suburban-Outdoors Neighborhood	n/a	48	Midday ambient sound level, low traffic volume
Residential Smoke Detector	See comment	85	Ref. at 10 ft. distance. NFPA 72, National Fire Alarm Code
Commercial Fire Alarm System	See comment	110	Max. at minimum distance, NFPA 72, National Fire Alarm Code
Inside SUV Vehicle	n/a	36	Center top dash level measurements, windows & doors closed, engine off
	n/a	47	Windows & doors closed, engine running
	n/a	60	Windows closed, driving 35 mph
Office Building Mechanical Space	n/a	70	Average of sound levels measured in an open floor plan mechanical space
Building AC Vent Fan	3	68	Vent fan operating at 850 RPM
Building AC Vent Fan	3	85	Vent fan operating at 850 RPM and guard housing was vibrating
Industrial Combustion Fan	3	80	Combustion fan operating at 1525 RPM
Computer Room	n/a	64	Room with 43 operating desktop sized computers. Center of room measurement. Room size (3.3 m W x 7.3 m L x 3.3 m H)
Fan–Smoke Ejector	3	92	Electric variable speed fan on max. setting 2200 RPM

The dBA values measured in this report do not fully account for all of the sound that may be experienced on a fireground, but is a sample of sounds that may be experienced. It also does not account for the level of sound lost by fire fighters who are properly wearing their personal protective equipment (PPE), i.e. helmets; ear flaps, hoods, and coat collars can reduce the amount of sound that a working fire fighter may hear. Results from these measurements of background and fire equipment sound levels demonstrate that the 95 dBA sound level from a PASS device alarm has the potential for being lost within the normal sounds generated while responders are operating on the fireground.

Figure 16. Photographs showing background sound level measurement tests.

4.0 RECOMMENDATIONS FOR FUTURE WORK

Sound data collected during this study demonstrates that sound levels can be muffled by changes in firefighter positions or the orientation of the alarming PASS device. Additionally, potential sound levels associated with fireground operations may be loud enough to challenge PASS device alarm sound performance. Additional research areas which could provide better characterization of the attenuation of PASS alarm signals include:

- Measure the sound attenuation (muffling) associated with blockage by furnishings and building contents.
- Measure the muffling associated with a range of different types of construction, walls and floor ceiling systems.
- Measure sound levels with multiple pieces of fire apparatus and fire fighting equipment running.
- Measure sound level attenuation caused by the proper use of PPE.

5.0 CONCLUSIONS

Data from this limited study demonstrate that PASS alarm sound levels can be muffled by personnel wearing typical firefighting ensembles when oriented in the test positions specified in NFPA 1982. Additionally, the data demonstrates that PASS alarm sound levels may be reduced by simply changing the unobstructed orientation of the PASS device. The findings suggest that the loss in PASS device amplitude due to orientation and muffling has the potential for reducing sound levels to a point that may make it more difficult for rescue personnel to locate an incapacitated firefighter.

Sound measurements presented in this report for PASS muffling and ambient background, fire fighting apparatus, and fire fighting equipment show that the 95 dBA PASS device alarm level may be seriously challenged by other sounds exposures on the fireground. The impact of muffling of the alarm signal and background sounds should be incorporated into performance metrics for PASS devices to insure the alarm signal are adequate to locate incapacitated fire fighters

6.0 ACKNOWLEDGEMENTS

Appreciation is extended to members of, Company 53, the NIST Fire Protection Group: Chief Thomas Rhodes, Chief Danny Baker, Capt. Mark Miller, Lt. Ivan Todd, and FF/EMT Eric Inkrote for their assistance in conducting this study. Appreciation is also to extended members of, Company 12, the College Park Volunteer Fire Department, Prince George's County, MD, particularly Deputy Chief Stephen Kerber and Firefighter Matthew Machala for their assistance in collecting apparatus sound data for this report.

The support of U.S. Fire Administration, especially project officer Brad Pabody is greatly appreciated.

7.0 REFERENCES

1. National Fire Protection Association, NFPA 1982 Standard on Personal Alert Safety Systems (PASS), 1982 Edition, Quincy, MA.
2. Federal Interagency Committee on Aviation Noise, "How do we Describe Aircraft Noise?" www.fican.org.
3. U.S. Department of Health and Human Services, "Criteria for a Recommended Standard, Occupational Noise Exposure, Revised Criteria 1998," Centers for Disease Control and Prevention, Ohio, 1998.
4. National Fire Protection Association, NFPA 1982 Standard on Personal Alert Safety Systems (PASS), 2007 Edition, Quincy, MA.
5. National Fire Protection Association, NFPA 72, National Fire Alarm Code, 2007 Edition, Quincy, MA.

www.ingramcontent.com/pod-product-compliance
Lightning Source LLC
Chambersburg PA
CBHW081822170526
45167CB00008B/3502